Cars

10 Things You
Should Know

Lewis Kingston is an award-winning journalist and author with a degree in motorsports engineering. He has worked for several prominent automotive titles and tested all manner of vehicles, from entry-level hatchbacks to flagship supercars. To further develop his experience and understanding, Lewis also regularly changes and works on his own cars, with his back catalogue including a twin-turbocharged Toyota Supra, a Dodge Charger, a Lancia Delta HF Integrale, a BMW M5, a Rover P5B Coupé, a Suzuki Swift Sport, and some fifty other classics, modern classics, and curios. As well as continuing to get his hands dirty in the workshop, and alongside his ongoing contributions to numerous motoring outlets, he is also working on both non-fiction and fiction book projects, including sequels to his self-published science fiction novel, *Prompt Excursion*.

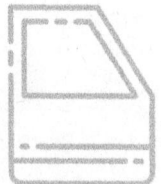

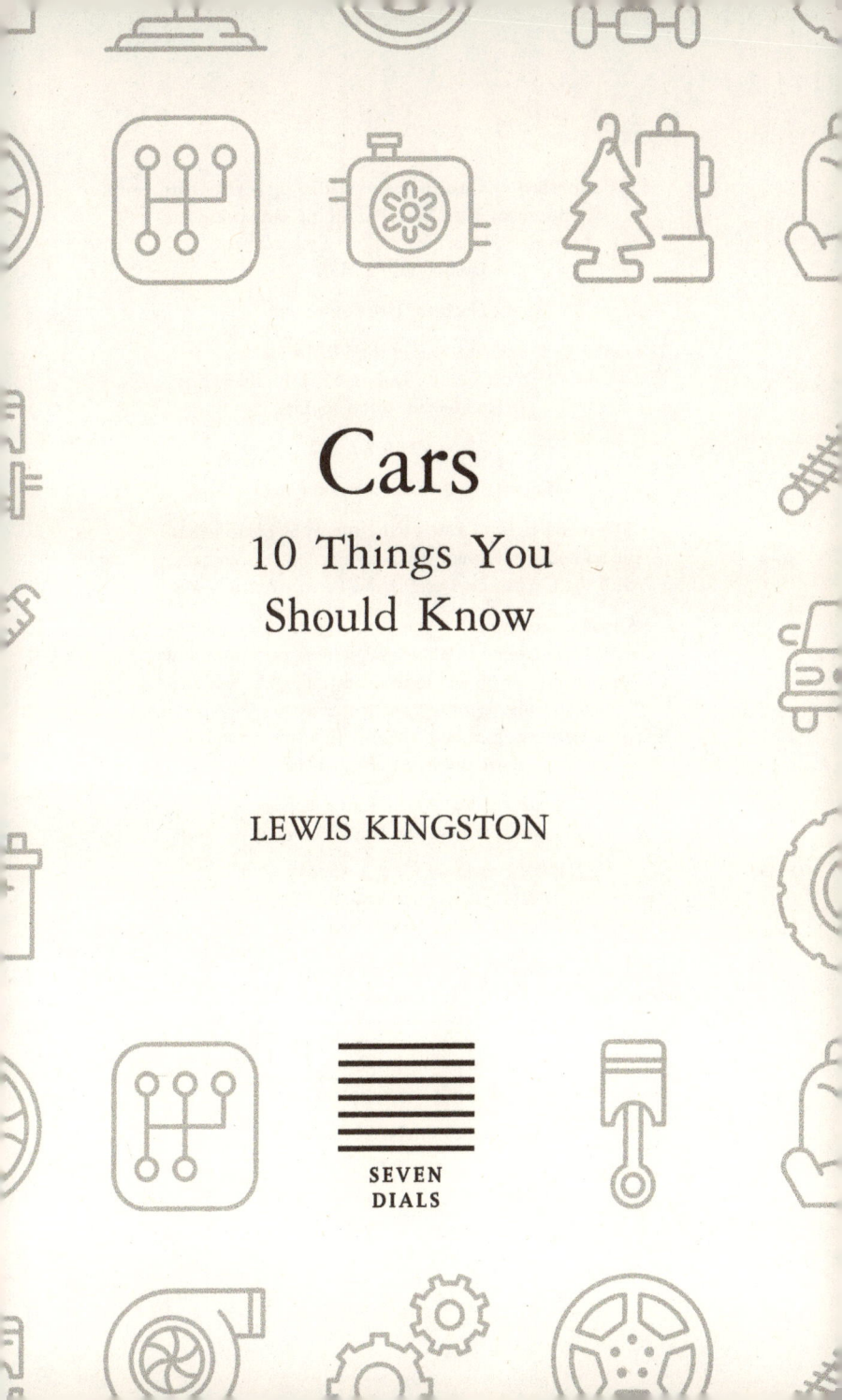

Cars

10 Things You
Should Know

LEWIS KINGSTON

SEVEN
DIALS

First published in Great Britain in 2025 by Seven Dials,
an imprint of The Orion Publishing Group Ltd
Carmelite House, 50 Victoria Embankment
London EC4Y 0DZ

An Hachette UK Company

The authorised representative in the EEA is Hachette Ireland,
8 Castlecourt Centre, Dublin 15, D15 XTP3,
Ireland (email: info@hbgi.ie)

1 3 5 7 9 10 8 6 4 2

A CIP catalogue record for this book is
available from the British Library.

ISBN (Hardback) 978 1 3996 3343 7
ISBN (eBook) 978 1 3996 3344 4
ISBN (Audio) 978 1 3996 3353 6

Typeset by Born Group
Printed and bound in Great Britain by Clays Ltd, Elcograf S.p.A.

www.orionbooks.co.uk

For Romana

Contents

Preface

I vividly remember the first time I drove a car. Okay, perhaps 'drove' is a bit of a stretch: I was four, couldn't reach the pedals, and it was on a dirt road. But, perched on the owner's lap, I took hold of the steering wheel and merrily swung it back and forth. The car responded in kind, its engine burbling away melodiously, as it bounced down the rough trail. The sensations, the sounds, the sights, were invigorating and exciting. And, among everything else flying through my mind, there were new and countless tempting possibilities: we could go to the end of the road, to the woods, or perhaps the next village, or the next town. And what, or who, might we discover there? Back then, admittedly, just finding an ice cream vendor appealed the most to me. But the memories, those inexorably intertwined with the car, would continue to accumulate: trips with my family, trips with my friends, the mistake of trying to clean dirt off my mum's Saab with wire wool, my dad surviving writing off his Ford Granada in spectacular style, being violently ill in the back of a Citroën 2CV, the flat floors of which definitely did us no favours that day, and the rest, all seared into my memory. And that was all before I even got into a car of my own.

I'm sure many of you, reading this now, have similar recollections. In some cases, those formative experiences may have planted seeds that bloomed into an interest in cars, which is why you've subsequently bought or been gifted this book. Alternatively, it could be that you're just getting started and would like to know more about what's out there in the land of cars, and how it came to be. And some of you, I know, will be on the receiving end of a present-giving error of judgement, and you find cars as enthralling as dishwashers or vacuum cleaners. Whatever the case, I hope that in the next ten chapters, assembled by drawing on some fifteen years' work in the automotive field, you find something that sparks or sates your curiosity. There is all manner to discover on this whistle-stop tour, after all, ranging from the astonishing scale at which cars are built, to the speeds they can achieve, and how they have influenced our society and surroundings. And so much, one way or another, that could prove significant to you, as the car and the world continues to evolve. But let's not sit here idling at the lights any longer: make yourself comfortable, and I hope you enjoy the trip.

From One to 1.5 Billion

France had a problem. The Seven Years' War, which effectively started in 1754 and lasted nine years, had ended in defeat. The country had battled to strengthen its position and expand its overseas territories, during what was arguably the first truly global conflict, but it instead suffered vast losses across the board. In the disastrous peace deal that followed, France ended up having to hand over a significant portion of its North American territory to its arch-enemy, Great Britain, significantly tarnishing and weakening its empire. As far as crises went, it was a massive one, and the shock waves rippled throughout the country. It couldn't be allowed to happen again, and France had to recover, to rebuild, to bring to bear an army that could protect the country, both at home and overseas. And after all the failings, and all the torrid mistakes and miscalculations that had been made, it was clear that a substantial overhaul of the military was required. Unbeknownst to those involved, the decisions that would follow would lead to the creation of an incredible machine, one that would reshape the world: the car.

Following the war, one aspect of the military that required improving was mobility. An army that could respond quickly, one that could better pick where it wanted to fight, or bolster its forces at short notice, was more likely to triumph. But it wasn't just soldiers that needed to get from A to B, at least not if you wanted them to be effective: cavalry support and logistics aside, such as food and ammunition, they needed artillery. And the big guns, such as those in the Vallière system that had been introduced in 1732, were no joke. The light four-pounder cannon, for example, reportedly weighed in at 1.2 tonnes with its carriage and needed four horses to drag it around. The heavier twelve-pounder gun? Some 2.5 tonnes and nine horses. This dependency on horses, which needed to be sourced, fed, rested, and cared for, added another stack of logistical challenges to the military's ever-lengthening lists. There were also some obvious problems with using horses on a battlefield, including their inability to pull the gun carriages in poor conditions, and their incompatibility with explosions.

There was one innovation, that had been growing in prominence, which posed a solution: the steam engine. You took water, heated it in a boiler, and the expanding steam created could drive a piston back and forth, producing useful mechanical action. Even though the industrial revolution had yet to start in earnest, the steam engine was already making waves. The first commercial engine had made its working debut in 1712 and, as the technology began to spread and develop, the industrial advantages became clear.

Steam engines were consistent, they didn't need to sleep, they didn't defecate everywhere, and they were relatively easy to use and maintain. Although steam engines were strictly stationary at this point, those in the military, tasked with fielding, employing and advancing its capabilities, began to envision ways they could employ this groundbreaking technology.

Among those who recognised one significant opportunity was Nicolas-Joseph Cugnot. He was born in France, served in the Austrian army as an engineer, lived through the Seven Years' War, and settled in Paris in 1763 after leaving the military. There, no doubt driven and informed by his experiences, he set about writing books about military fortifications and continued to experiment with the hot new thing: steam power. In particular, he was intent on working out how to employ a steam engine to propel a carriage, removing the need for reliance on horses. Cugnot had to solve lots of problems along the way, including figuring out how to convert the back-and-forth motion created by steam engines of the time into something that could rotate a wheel. Eventually, he managed to draft a workable proposal and sent it off to the army for their consideration in 1769. It ticked the forward-looking boxes, being innovative and potentially ideal for towing artillery, and a prototype was duly commissioned.

Later that year, Cugnot unveiled the first workable self-propelled land vehicle, called the *fardier à vapeur*: a steam dray, a dray being a strong wagon. But the machine, which could be started, move under its own power, stop, steer, and carry passengers, was more than just a wagon, it was the first car. His invention had three wheels, with the single front steering wheel being driven by a two-cylinder steam engine, based on an earlier design by German engineer Jakob Leupold, with the boiler hanging out in front. Unsurprisingly, it bore a passing resemblance to a carriage being pulled by a horse, with the horse having been replaced by a hissing, smoking, steaming contraption. And, reputedly, with four people aboard, it erratically hammered its way through the grounds of the Paris Arsenal during its first trial and was estimated to be capable of reaching up to 2.5mph (four km/h). Only on paper, though, because it could only run for roughly fifteen minutes before it exhausted its steam and ground to a halt. In one report, it was noted to have covered less than one mile in an hour. In short, those spectating could have walked over three times the distance in the same amount of time.

This didn't put off the observers, though: this was only a concept, after all, and it was never going to be perfect out of the gate, and the production of a larger, more powerful version was authorised. This scaled-up beast, which clocked in at 2.5 tonnes unloaded, was designed to pull up to 5 tonnes at similar speeds. Once it had built up steam, the improved wagon was also reported to be capable of running non-stop for an hour and fifteen minutes. Several problems persisted, however, including slow speed, instability due to its three-wheeled layout and poor weight distribution, and many remained dubious of the real-world potential of such a bizarre device. Remember: this was more than thirty years before the first steam train even existed, so it was a pretty wild idea to get to grips with. The ever-shifting political landscape also meant that funding for the steam carriage, and its support, soon disappeared, leaving it dead in the water.

Eventually, a sheet would be pulled over Cugnot's carriages. The first would disappear into the mists of time and the second ended up in the Musée des Arts in 1801, where it resides to this day. That all said, the concept of a wheeled vehicle that could move under its own power, even in the early 1800s, wasn't a new one. The idea had been conceived as early as the eighth century BC, some 2,600 years prior, and detailed in the Ancient Greek poem *The Iliad*, where Homer described wheeled tripods that moved around autonomously. A fifteenth-century precursor to the car was also penned by Leonardo da Vinci. The famed Italian artist, theorist, and engineer, outlined a self-propelled cart, powered by coiled springs, which had no driver. Instead, its brake would be released remotely, and it would follow a pre-set path with its cargo, much like Homer's automated tripods. Myriad other ideas would follow, too, some employing steam, some clockwork, some hopes and dreams, but Cugnot's efforts were ultimately the first feasible, functional, and well-documented vehicles that you could sit in, stoke up, and take for a drive, without the wind or a hill in your favour.

Admittedly, a 2.5-tonne juggernaut wouldn't become the transport of choice for a couple of hundred years or so, and we were still some way away from the arrival of what is generally accepted as being the first passenger car. After Cugnot's invention fell by the wayside, others, including British inventor and engineer Richard Trevithick, would pick up the steam baton and run with it, developing more practical and capable technology. In 1801, Trevithick unveiled a road-going steam vehicle, nicknamed the Puffing Devil, and demonstrated it by carrying passengers around Camborne, Cornwall. And while Trevithick's engineering advancements would set the stage for the first steam locomotive in 1804, unleashing another torrent of industrial development, steam power remained impractical for use in smaller vehicles. The boiler needed to produce steam was heavy, as was the water and fuel you had to bring along with you, and managing everything properly, to avoid the whole thing burning to the ground or exploding in your face, was no mean feat. Then there was the small matter of convenience: imagine going out to your car, turning the key, and then having to wait half an hour before it could move under its own power. These problems would eventually be tackled but, at the outset, they were stumbling blocks for early inventors and adopters alike.

Fortunately, before long, a new and more suitable motive power rolled noisily into town: the internal combustion engine. Unlike steam engines, which burnt their fuel externally, these engines burnt their fuel inside, making them smaller, lighter, and less complicated. And, as they became more practical and serviceable, they opened the floodgates for the development of horseless carriages. Many contributed to their development, but Swiss inventor François Isaac de Rivaz is the first person documented as putting one on a small carriage, in 1807, creating the internal-combustion-powered vehicle, one which used hydrogen gas as its fuel. Steam, electric, and even compressed air propulsion remained in development but, as internal combustion engine technology surged ahead, and we'll look at more of that later, it became the power source of choice for many. Including, among the many striving to make something useful and marketable, German engineer Karl Benz. Yes, you're on the right track if you think that surname sounds familiar. One more thing: he later often wrote his name as Carl, so we'll use that when required from here on out.

Benz had worked for several engineering firms before setting up a company producing gas-powered engines, developed by himself, in 1883. But engines weren't the heart of his focus: he wanted to build a motor car, to realise this long-standing concept of a truly horseless carriage, and used his new industrial foothold to set about developing the technologies and designs required to make it happen. By October 1885, he had finally completed his project, which he filed a patent application for on 29 January 1886: Patent No. 37435, for a motorised vehicle powered by a gas engine. To many, this date marks the birth of the automobile, in the form of the Benz Patent-Motorwagen. And while Benz's car only had three wheels, because he'd not yet figured out how to effectively steer a car with four wheels, the 0.74bhp (0.55kW) Model 1 established a template that others would follow, employing a chassis, a lightweight internal combustion engine, fuelled by a petroleum derivative, a transmission, brakes, cooling system, steering gear, and more. It wasn't a one-off prototype, either, unlike some others: Benz would build twenty-five, before forging ahead with a more advanced four-wheel successor, called the Velocipede, also known as the Velo, in 1894.

It's worth noting that Benz wasn't the first to submit a patent for a car, though: that was American inventor George B. Selden, in 1879, but revisions and delays meant the four-wheeled concept wasn't approved until 1895. And Benz wasn't the first to build a car, as we've established. Nor was he the first to build a road vehicle in series, as large, innovative steam carriages had already been built in number by French pioneer Amédée-Ernest Bollée, starting in 1878. But Benz was the first to persist with the idea of a smaller passenger car powered by internal combustion, something that could be privately owned and managed, and to prototype it, to develop it, to build it, to test it, to advance it further, to market it, to bring a usable and viable product to the world, setting the scene for decades to come. The debut of the Patent-Motorwagen was unquestionably a significant event, and Benz's achievements shouldn't be understated, but countless others, including Siegfried Marcus, Étienne Lenoir, Louis François René Panhard, Émile Constant Levassor, Nicolaus Otto, and Édouard Delamare-Deboutteville, all worked tirelessly in these formative years to pave the way for the cars that we know today.

And now? Well, as you've probably guessed from the title of this chapter, unless you're reading this after it has fallen through a rift in time, there are estimated to be some 1.5 billion vehicles rolling around on the planet. And, as you'll soon find out, that number, for better or for worse, grows larger with every passing minute.

How the Car Changed
Our World

Let's go for a walk. It might seem counterintuitive, considering the subject of this book, but walking was originally all we had. If you wanted to get somewhere, be it for work, to meet someone, to gather food, to simply survive, you had to walk. The process of putting one foot in front of another, literally and figuratively, was the linchpin of our lives. Even once we'd started using horses and other animals for travel, walking remained the common and free way to get around on land. And life wasn't the only thing that revolved around walking: our villages, our towns, our cities, and everything else, was designed to allow you to get by using just your own two feet. This had its advantages, including keeping people healthy and building stronger, more engaged communities. But it also had its drawbacks, as many people were effectively confined to one area, with travel further afield often being too expensive, and too time-consuming, reducing their opportunity to explore the world and to develop.

The debut of Carl Benz's Patent-Motorwagen didn't, however, immediately cause a seismic change in how we got around. As with Cugnot's earlier steam carriage, and as was the case with the other production models and concepts drifting around, the public and investors alike were sceptical, and sometimes scared, by the strange new concept: these machines were noisy, noxious, dangerous, expensive, and their distance-covering capabilities seemed limited. Instead, it was bicycles, in terms of mass mobility, that were making the most dramatic impact in the late nineteenth century. At the time, the penny-farthing, which had one large, pedalled wheel, and a smaller trailing wheel, was a popular choice, but also precarious and perilous. But in 1885, a new generation of bicycle arrived that offered similar thrills and escapism, but without the spills: the safety bicycle. One of the most commercially significant, the Rover Safety Bicycle, launched by British inventor John Kemp Starley, set the template for the machines we ride today. Its wheels were the same size, unlike those on the tall penny-farthing, and it had a chain-driven rear wheel. It wasn't cheap, costing £22 at a time when your annual wage might be £60, but it was at least half the price of a good horse, a tad more manageable and didn't require feeding, grooming, exercising, stabling, or veterinary care.

As the popularity of cycling grew, new manufacturers sprang up to satisfy the demand, in turn driving down costs and making bicycles even more accessible. Personal mobility was now more affordable and viable, allowing people to travel to new homes, towns, and cities, or to begin commuting to jobs that were further afield. This new method of readily accessible travel had a significant effect on class mobility, as people could now travel beyond their previous means, granting them access to new careers and opportunities. It also had a significant impact on gender mobility: in the Victorian era, a woman was expected to find a partner, marry, raise children, and carry out their domestic duties at home, and that was their lot. Which, of course, was patently nonsense. But the bicycle, and the independent mobility they offered, provided them with a way to break from that norm, to justify being outside, allowing them to explore, to socialise, to gather, and to expedite the development of their freedom and rights.

Bicycles also helped the car itself come to life. A great deal of the parts used in early cars were either directly borrowed or derived from those used in bicycles. Gears, chains, bearings, braking parts, steel frames, wheels, and even names, were lifted from the bicycle industry to enable cars to be developed and built. Furthermore, it was cyclists that helped smooth the way for cars, as they were vigorously pushing for the construction of new roads, to replace rough, unsurfaced routes in cities and the surrounding areas with better infrastructure that was safer, cleaner, and easier to cycle along. In England, this included organisations such as the Bicycle Touring Club, later renamed the Cyclists' Touring Club, while in America and Canada, substantial lobbying was carried out by the cycling-focused Good Roads Movement. The latter, in particular, brought farmers into the fray more prominently, expressing and demonstrating how advantageous it would be for them if the roads were better, how it could save them time and money, and how such infrastructure needed to be viewed as a tool. In poor condition, it wasn't going to do them much good, and if revamped and expanded, it would make moving around easier for everyone.

But in 1888, the wheels started to turn a little in favour of the car. Benz had been working away on his Patent-Motorwagen, and had developed a much-improved version called the Model 3, but no real success had been struck. It wasn't entirely difficult to understand why people weren't flocking from their horses and bicycles to it: the car was still slow, complicated, unreliable, expensive, and there was no proof that it could go very far. Compounding the problem, Benz just wasn't much one for marketing, or business operation in general, and even he was seemingly beginning to doubt the viability of his creation. His wife, Bertha Benz, on the other hand, was a firm believer that what he had built was of great significance. She also had a vested interest in her husband's success as, prior to their marriage, Bertha had her dowry paid out early so she could settle Carl's business debts and allow him to pursue his car-related ventures in earnest. Her faith in her husband's project was endless and, despite the setbacks, she remained optimistic and strove to find a way out, to trigger that lightning-in-a-bottle moment they needed.

On 5 August 1888, Bertha Benz did something unheard of: she took her husband's Model 3 Patent-Motorwagen, without his knowledge, and drove it from Mannheim to Pforzheim, with her two sons as passengers. It took them over twelve hours to cover some 60 miles (100 kilometres), through southwest Germany, and during the pioneering trip she solved countless problems. Bertha found a chemist selling the petroleum solvent that the engine needed to run, inadvertently turning the shop into the world's first petrol station in the process, used her hat pin to clean a blocked fuel line, tore material from her garter to insulate a shorting wire, found a blacksmith to mend a drive chain, and came up with recommendations such as adding a gear in the transmission designed to make it easier to drive the car up hills. Once they had arrived at their destination, Bertha sent a telegram to Carl to inform him that his car had completed its first long-distance trip, then drove back home a few days later. Thanks to her bravery, and her ambition, the capability and practicality of Carl's creation had been demonstrated. Here, in self-propelled form, was a new kind of freedom; one that could let anyone who could afford it roam far and wide, independently, privately, and spontaneously.

The trip itself generated much in the way of publicity, in part because many of those passed by the Patent-Motorwagen had never seen such a smoking carriage that was seemingly being moved by nothing at all. Coupled with the influx of interest that followed, and the confidence reinstated in Carl by Bertha's demonstration of the car's abilities, a fresh flame was lit under Benz's company. By 1901, he would have built over 1,200 of his later four-wheeled Velo alone, and other companies had sprung up to establish a foothold in the market for cars, including Oldsmobile, Fiat, Renault, and Packard. The number of cars in the wild remained small, though. In the UK, for example, there were only some 800 cars on the road in 1900. In modern terms, that's not even enough to fill a large multi-storey car park. But as the 1900s started rolling in earnest, as production surged, technology marched on, and costs started to tumble, the car started shifting from a low-volume toy for the wealthy into a more attainable and accessible mass-produced mode of transport. Changes in society were also facilitating the wider adoption of the car, including expanding industrialisation that was creating new jobs and increasing earnings. It was also becoming clear that cars could replace horses, in terms of reliable long-distance travel, as they became more affordable and more dependable. Cars were continuing to improve, after all, whereas horses were not. Horses also posed an increasingly significant issue that urgently needed to be tackled: the vast number required to move everyone around, to allow cities to operate, was literally flooding the streets with urine and faeces, causing endless health issues. In London alone, over 300,000 horses were slogging their way through the streets every day.

For some countries, the prosperous years that followed World War One led to another vast upswing in the number of vehicles on the road. By 1920, there were 187,000 private cars alone on the road in the UK, 234 times as many as there had been just 20 years prior. In the United States, there were now some 8 million, up from 8,000, a 1,000-fold increase. After World War Two, the figures surged even more significantly: by 1950, there were 4 million vehicles registered in the UK, including light goods vehicles, and over 40 million in the United States. Further to this, as more people wanted to drive, bought cars and took to the road, the car started bringing about changes to the world. Much like the bicycle, cars were used to promote causes, including gender- and race-related matters, allowing people to spread their messages and raise awareness, further enabling social mobility. Ongoing development and rising car ownership also drove expansion of the road networks, with public road coverage growing from 2.3 million to 3.3 million miles (3.7 million to 5.3 million kilometres) in America between 1900 and 1950. Significantly, before 1900, only around 4 per cent of those roads had any kind of surfacing. By 1950? Some 23.5 per cent. Increasingly, as car ownership and usage continued to grow, out-of-town malls and shopping centres began to appear, along with service stations, car parks, and ever-expanding urban sprawl.

Since then, the numbers have only climbed. But the pitfalls of cars were and remain numerous. The environmental impact is significant, whether in terms of engine emissions, particulates from tyres and braking components, congestion, or noise, and affects not only us but the world in which we live. Let's not overlook the negative environmental effects of building the cars themselves either: the raw materials consumed, the emissions produced by shipping them around the world, and the production of the fuels and batteries required to power them. Then there are the collisions and the incidents. The World Health Organization, as of December 2023, estimated that 1.19 million people died in road traffic collisions and incidents that year, while between twenty and fifty million people suffered injuries. Look also at where we live and work: how much of it is designed to support the car, and what are the societal and environmental problems that that causes? Technology may offer some solutions, yes, along with changes in how we regulate and use cars, but whichever way you cut it, the convenience and freedom the car offers commands a hefty price.

An Introduction to
Engines

It's 8 a.m. on a cold winter's morning. The kettle boils away merrily, the steam condensing on the window, as you set about your breakfast, ahead of leaving for work. Eventually, with a sigh, you unlatch the door, step out into the fresh, crisp air, and unlock your car. It's all routine: get up, get ready, start the car, commute to work, do your job, return home, rinse and repeat. And cars, for the most part, are predictable, reliable, and go about their task of moving you around the world with minimal fuss. Which makes it easy to forget that what you're driving, what you're witnessing and engaging with every day, is a finely orchestrated symphony of countless complex systems working together, the result of hundreds of years of development, and the incremental but relentless improvements that come about through building more than 1.5 billion vehicles.

At the heart of any car is its engine or, increasingly so, its electric motor. It's what produces the power to move the car, and what provides the feeling of acceleration that many enjoy. But most cars, even today, still rely on an internal combustion engine. At the most basic level, it's a machine that converts stored chemical energy into mechanical energy. In this case, that's achieved by burning fuel with oxygen in an enclosed space. And, if you've ever heated up a microwave meal, and watched the plastic film bulge upwards, you already know what happens when you heat gases and liquids: they expand. It's this expansion that an internal combustion engine harnesses to create the force that gets the car moving on down the road. However, it's also the engine, often the linchpin of people's love affair with cars, for the noises and sensations it produces, which remains the most mysterious element for many.

After all, if you lift a car's bonnet, you'll typically just find a daunting mass of metal and pipework, or a slab of detail-obscuring black formed plastic. You'd be justified, as a result, to just deem it one of those magical things, shut the bonnet, and carry on with your life. But, if you pare it back, the basic design and operation of an engine is relatively straightforward. Let's cut through the clutter by building a basic engine, as a demonstration. We're going to need a chamber in which we can burn fuel, in this case petrol, so we'll start with a hollow metal tube. This is going to be our cylinder. We're also going to need something to harness the power from expansion of the burning fuel-air mixture, so we'll put a piston, a small disc of metal, inside the cylinder. We'll seal the top end of the cylinder so that, when the mixture starts to burn, the only way it'll be able to expand will be by pushing the piston down the cylinder. Now, if you put the piston in this cylinder, and fill it with fuel and air, and ignite it, you'll just fire the piston out of the open end of the cylinder. Which, while it might look cool, isn't useful for our needs. And the straight, linear motion of the piston isn't much good if you want to turn a wheel, either. We need to create rotation, and to do it repeatedly.

If you've ever pedalled a bicycle, you've already experienced what we're going to add next: a crankshaft. When you're cycling, your feet press down on the pedals, which turns the bicycle's crankset. This assembly, which consists of crank arms and sprockets, converts the up-and-down motion of your legs and feet into rotation. This then drives the chain, turning the wheel, propelling you along. We want our engine to be able to do the same, so we'll put a crankshaft at the bottom of our engine's cylinder. We don't want to have to pedal, though, so we'll connect our piston to where a pedal would have otherwise attached, using a component called a connecting rod. The piston can pivot on the top of it, and the connecting rod can rotate around the crankshaft, so everything can move freely.

We then turn the crankshaft, so the piston is at the top of the cylinder, introduce some fuel and air, ignite it, and stand back. The mixture inside the engine starts to burn, expands, and pushes the piston down the cylinder. This moves the connecting rod, turning the crankshaft, and voilà! We've created a machine that converts stored chemical energy into useful mechanical energy. At least, that is, briefly: the energy in the hot fuel-air mixture soon dissipates, and the piston slows to a halt at the bottom of its travel. There's nothing to push it back up the cylinder, as there's no more energy in the system, so we can forgive it for only doing its job once. What we need is a flywheel, a heavy metal disc bolted to the crankshaft, that stores rotational energy. It's the mechanical equivalent of a capacitor in an electrical circuit and, once the crankshaft starts turning, it'll store the energy needed to drive the piston upwards. Not only will this keep the engine running, but it'll also smooth out its operation and make it less likely to shake our workshop to pieces.

You may have noticed a stumbling block, however: what do we do now that the cylinder is full of spent gases and waste products? We can't just ignore it, as if we want the engine to keep going, we're going to need to get the waste out and fresh air and fuel back in. We can't just go 'Expecto petroleum' and, voilà, magically refill our cylinder with a mixture that's ready to go. Instead, we need to be able to open and close passageways leading into the cylinder, so a fresh fuel-air charge can get in, and waste can get out, allowing the engine to fire more than once. In most conventional automotive engines, and where things get a little complicated, this is the job of the cylinder head and valve train.

Let's just refresh our memory: we have our piston, connecting rod, and crankshaft, which together are called the rotating assembly. These all sit inside the engine block, a metal casting which contains the cylinder and provides a home for the crankshaft. To seal the cylinder and create a place for combustion to take place, we'll employ what's called a cylinder head. This is typically an intricate piece of cast metal, inside which are passageways. These provide a way for fresh air to get into the cylinder, and for exhaust gases to get out. Now, these can't remain open all the time, otherwise we'd not be able to make use of the pressure generated during combustion. It would be like trying to puff your cheeks up without closing your lips: you'll blow a lot of air out, make a lot of noise, and expend lots of energy, but nothing useful will happen.

To regulate the flow through the cylinder head's intake and exhaust ports, we need valves. Think of them like your fingertips on the holes of a flute. You want to change the note, to control the flow of air, so you cover or uncover the tone holes. A valve in an engine does a similar thing, opening or closing when required to allow us to control what's happening. These valves look like the upper half of a spinning top, with a long stem, and we'll hold them closed with springs.

This will stop them, for one thing, from falling into the engine. With the valves keeping the ports in the head closed, we've got our cylinder ready for combustion, but how do we open them? Well, we'll regulate them with camshafts, which are metal shafts with specially shaped sections on them called lobes. Put an egg on its side on a table, and gently push your finger against the top of the egg, and you'll see how these work: rotate the egg and you'll see that your finger, and hand, will move back and forth as your fingertip follows the egg's profile. The lobe on a cam does the same, transferring motion to the tip of the valve stem, and controlling how and when the valve opens.

This allows us to let a fresh mixture into our cylinder, to seal it for combustion, then to let exhaust gases out when we're done. And, neatly, we can drive the camshaft with the crank, through a belt, chain, or series of gears. This automates the operation of the valvetrain, including the cams and valves, and keep it in time with the crankshaft rotation and piston movement, so everything happens when it needs to. And this complete process, the one used in most cars around the world, is called the four-stroke cycle. We start off with the piston at the top of the cylinder, a position called top dead centre, and all the valves are closed. The intake stroke kicks off the proceedings: the intake valve in the cylinder head opens, and the piston starts to move down, drawing fresh air and fuel into the cylinder. At the bottom of the piston's stroke, called bottom dead centre, the intake valve closes, and the piston starts to move upwards. This, the compression stroke, causes the mixture to get squished, which helps us extract more energy from it. The mixture is then ignited, starting the combustion stroke, and the rapid expansion of the gases in the cylinder push the piston downwards, turning the crank. As the piston starts to travel back up again, the exhaust valve opens, allowing waste from combustion to be pushed out. And, once the piston nears the top of the cylinder, the exhaust valve closes again, and the cycle starts again. Intake, compression, combustion, exhaust, over and over. Or, as the saying goes, suck, squeeze, bang, blow. No giggling, please.

About that fuel and spark, though. You're probably wondering, if you're not familiar with engines, how we're delivering that fuel into the cylinder, and how we're creating the spark that's needed to start combustion. The spark isn't too tricky as, in a petrol engine, we can place a spark plug between the valves in the cylinder head. It's like the igniter on a gas stove but, instead of you pressing the button to trigger it, it's initiated electronically or by mechanical actuation. At the right moment, an electrically generated spark jumps across an air gap in the spark plug, igniting the swirling fuel-air mixture around it, starting combustion. And, in mainstream modern engines, fuel is provided by electronically controlled injectors that spray fine mists of petrol into the intake manifold, or directly into the cylinder. Typically, an electronic engine management system, aided by numerous sensors, monitors the engine and regulates its systems to make sure it runs efficiently.

All told, once your single-cylinder, spark-ignition, four-stroke petrol engine is idling at 500 crankshaft revolutions per minute, a combustion cycle will take place just over four times every second. There's typically a lot more to car engines, though, with most having multiple cylinders, usually from two to twelve, in various arrangements, with the most common being the straight or vee-type engines. There are engines with two, three, four, or five valves per cylinder, multiple camshafts, forced induction to boost power, variable valve timing and lift systems, and more. Then you've got lubrication, cooling, emission controls, diesels, two-stroke engines, Wankel engines, no giggling again please, and the rest. And that's just scratching the surface when it comes to engines, and how a car gets down the road. But, like turning the ignition key of your car in the morning, it's a start.

The Science of Speed

It is 1894 and you have done something remarkable: you have built a functional and reliable petrol engine. But you don't want to stop there. Taking engineering concepts and ideas that you have read about, you have set about building your own car. In your garage, you put together a simple steel chassis, to serve as the car's backbone. At the front, you fit an axle, attached to the chassis with big leaf springs borrowed from a cart, with two wheels that can be steered. At the back, you place another axle, also suspended with springs to stop the car bouncing around wildly, with two wheels that are going to be driven by your engine. Between the two driven wheels, you install a differential. You heard about the concept from the cotton-making industry, where it's used in some of the machines, and have seen other car makers use it, so you think it might be useful: in a straight line, the wheels turn at the same speed but, to go around a corner, they need to rotate at different speeds, as one wheel will travel further. If they were joined by one shaft, the wheel on the inside of the corner would be forced to spin, while the outer wheel could drag, causing wear and making the car harder to drive. The gears in the differential, as the name suggests, allow the wheels to be powered but also to rotate at different speeds, and you hope it'll make your car quicker and easier to handle through the corners.

You'll need to stop as well as go, so you adopt the braking system you've seen on some bicycles: the spoon brake. In your car, it's just a lever that brings wooden blocks, with leather braking material on them, into contact with the wheels, slowing them down. That's the chassis, steering, suspension, and braking, all sorted. Next, you set your engine in the frame, at the front of the car, echoing the design of an 1878 steam carriage by Amédée Bollée that you'd studied. But, because it's early days, your engine doesn't make much power or torque, a twisting force, limiting its ability to turn your car's wheels. Your engine can also only spin at so many revolutions per minute (rpm), restricting the speed at which you can turn the wheels, slowing your ambitious hopes down. To clear this hurdle, you bolt a gearbox to the engine. In the same way that gears on a modern bicycle help the rider, the gearbox provides different gear ratios that allow the engine to accelerate the car to different speeds without having to struggle, improving your car's performance and efficiency. And, to momentarily separate the engine from the gearbox, so you can change gear or bring the car to a stop without stalling the engine, you fit a clutch. Power from the engine's crankshaft goes through the clutch, into the gearbox, then through a shaft to the differential, out to the wheels, and down to the ground.

Now, you can move on to wrapping things up. You don't want the engine to get too hot, as it would run poorly or turn into a pile of molten metal, so you adopt a simple cooling system that uses water to transfer heat from the engine into a radiator, keeping temperatures in check. Some basic bodywork, some seats, controls, and a fuel tank aside, your car is ready to roll. But how to prove its capabilities, to demonstrate its reliability, and to publicise it? You can't just hop online and hope for a viral social media post, as the internet won't exist for almost another century. #fail. Fortunately, correspondence from a friend gives you the answer: an event, organised by the French newspaper *Le Petit Journal*, called *Concours du 'Petit Journal' Les Voitures sans Chevaux*. A competition for horseless carriages. It is not the first organised race between self-propelled vehicles, that being at least as early as 1867 between two steam carriages, but it is the first significant competition involving cars. And, at this event, on 22 July 1894, a total of twenty-one cars will take place in the first organised trial of cars, driving from Paris to Rouen, a distance of 78 miles (126 kilometres).

It was a reliability test, first and foremost, a demonstration of what this new class of machines could do, and a chance for the press and public alike to see them in action. It also established a template of cars in competition, against the course, the elements, each other, and the clock, that would soon take hold as a source of great entertainment and excitement. Dedicated racing circuits and racing series would follow, include the first Grand Prix, which took place in 1906. But these increasingly popular competitive races and motorsports events didn't just deliver thrills and spills: they helped to push the technological boundaries forward. Not only in terms of performance, but in design, reliability, durability, serviceability, materials usage and safety too, with many concepts later trickling down into the mainstream automotive world, benefitting those driving their cars on the road, not just on the track.

It is now 19 April 2005, some 110 years after the Paris-Rouen competition, at the Ehra-Lessien test facility in Lower Saxony, Germany. The site, which features a 5.5-mile-long straight (8.8 kilometres), is around 240 miles (386 kilometres) from where Bertha Benz started her fabled road trip in 1888. But on this day, idling and surrounded by engineers, is not a single-digit horsepower car that could be outrun by a human, like Benz's or your creation: instead, it is a new Bugatti Veyron 16.4. The elegant, low-slung coupé packs 987bhp (736kW) and, as it rockets down the straight later in the day, it will establish a new production car speed record of 253mph (407km/h). At that speed, the luxury Veyron, which comes with a warranty, and meets all the regulations for different markets, from crash protection to emissions, is covering 113.1 metres (371 feet) every second. In doing so, it eclipses the long-standing record set by the McLaren F1 supercar of 1992, which could easily exceed 200mph (321.87km/h) in production trim, and achieved a record-setting 240.1mph (386.40km/h) derestricted in 1998, cementing the Veyron's position as one of the most significant cars of modern times.

One of the technologies employed by the Veyron, which helps it to produce so much power, to achieve its speed and accelerate from 0–62mph (0–100km/h) in 2.5 seconds, is turbocharging. An engine is effectively an air pump, and increasing its ability to move air will increase its output. More air going into each cylinder during the intake stroke means more oxygen is available, meaning more fuel can be burnt, making more power. But, even if you make the intake, the exhaust, the cylinder head, and all the other elements of the engine flow and work to their optimum, there's a limit to what's possible with natural aspiration, the engine breathing on its own. It's like asking you to take deeper and deeper breaths: there comes a point when you simply cannot inhale or exhale any more air. What allows you to overcome this limitation, in an engine at least, is forced induction: a compressor, driven electrically, mechanically, or by waste exhaust gas, which raises the pressure in the engine's intake, meaning more oxygen-containing air gets into the cylinders. Supercharging is when this compressor is driven by mechanical or electrical means, whereas turbocharging is when the compressor is spooled by exhaust gases. Think of a turbocharger like two hairdryers bolted together. One gets hot, fast-flowing exhaust gases pumped over its turbine, turning a shaft, which rotates a compressor wheel in the other, drawing in fresh air and compressing it, before it is fed into the engine.

In the Bugatti's case, its petrol engine has sixteen cylinders, in a W configuration, displacing eight litres, fed by four turbochargers. But power isn't the only element that's essential if you want to go fast: aerodynamics, tyres, gearing, and the rest, must all be perfected to allow a car to reach such heady speeds. Not that Bugatti stopped at 253mph (407km/h). In 2010, a Veyron Super Sport reached 267.85mph (431.06km/h). Then, when the Veyron's successor arrived, another significant benchmark was reached: in 2019, the new 1,578bhp (1,117kW) Chiron Super Sport 300+, as the designation suggested, reached 304.77mph (490.48km/h). It's not just production combustion-powered cars that are capable of immense speeds, though: the all-electric Rimac Nevera, which made its debut in 2022, packs an astounding 1,888bhp (1,408kW) and can get from 0–62mph (0–100km/h) in just 1.81 seconds. Flat out, it can reach 256mph (412km/h). The Aspark Owl, which entered production in 2020, goes one better, literally, clipping 257mph (413km/h).

However, these records pale in comparison to what's capable, for a wheeled vehicle, when the gloves come off and speed is the only consideration. Aerodynamic, ground-hugging, dart-shaped cars, with their wheels driven by everything from piston engines to gas turbines, frequently tear their way across the famous Bonneville Salt Flats at speeds far beyond the reach of even the mightiest production cars. Plus, if you do away with the complication of driving the wheels, and just use thrust alone? Well, you're into the land of the outright and ultimate speed record for a car, currently held by the twin-turbofan-powered ThrustSSC: it looked like a fighter aircraft without wings, and went like one too, achieving an average speed of 763.04mph (1,227.99km/h) over 1 mile (1.6 kilometres) in 1997. That's more than three times the top speed of a Veyron. And, yes, ThrustSSC exceeded the sound barrier, establishing the first supersonic record, streaking across America's Black Rock Desert at Mach 1.02.

While you and your car may benefit from the developments that the pursuit of speed delivers, driving fast, and experiencing it for yourself, might seem unfeasible or outright impossible. There are, however, opportunities for you to safely and legally experience higher speeds. Depending on where you are in the world, you should be able to access dedicated track days, drag strips, high-performance driving tuition, or even standing-mile events, along with instructors who will help improve your driving and allow you to better understand your car's capabilities. Not that you need to go fast to have fun: simply getting a car smoothly and neatly down a road can be satisfying and enjoyable in its own right. But if you want to sample something that feels faster, without immediately losing your licence or spending a fortune, try a classic car. A modern supercar will feel like it's barely moving at the legal limit, whereas an old roadster will often feel like it's doing twice the indicated speed, its pared-back nature amplifying the sights, sounds and sensations that otherwise make motoring so compelling. And when you want to go even faster, to find out what your car can really do, book some track time and let that engine sing in earnest.

5

Mind-boggling
Manufacturing

It'll take you about three seconds to read this sentence. During that time, for all intents and purposes, a new Toyota will have rolled off the production line. But, I hear you cry, the actual length of a year on Earth is 365.2422 days, meaning there are 31,556,926 seconds in a year, and that would mean Toyota is building over ten million cars in that time. Yes, that's exactly what it's doing: churning out the equivalent of one new car every three seconds, every hour, every day, seven days a week, every year. In 2023, for example, worldwide Toyota production, including cars from its luxury Lexus division, totalled 10,033,171 cars. But wait, there's more: if you include its subsidiary brands, Daihatsu and Hino, that tally rises to 11,517,622. These are not anomalous results, either, with the company's global production being close to, or exceeding, ten million vehicles annually for over a decade.

Producing this number of vehicles is no mean feat and requires a substantial amount of infrastructure. In Japan alone, Toyota has sixteen different factories, eight of which assemble cars. It also has fourteen plants in North America, three in Latin America, six in Europe, including one in the United Kingdom, four in Africa, and twenty-seven in the Asia-Pacific region.

The company, as well as producing vehicles on every continent, also sells and supports its products in over 170 countries. And none of these facilities, factories, hubs or dealerships run themselves, not yet at least, and Toyota employs over 380,000 people to keep its operations moving. Ensuring these facilities remain supported and supplied is a whole other story, on the business and employment front: in Japan alone, over 60,000 suppliers work with the manufacturer to provide everything required to keep the plants humming along, from production line support to parts, generating countless opportunities and jobs.

While Toyota is often at the top of the pile, it isn't the only brand delivering millions of new cars each year. The Volkswagen Group, which includes Volkswagen, Audi, Skoda, Seat, Cupra, Audi, and Porsche, produced 9.3 million cars in 2023. Stellantis? Across its many brands, including Alfa Romeo, Citroën, Dodge, Fiat, Jeep, Lancia, Peugeot, and Vauxhall, it shipped a total of 6.2 million vehicles. General Motors, which includes Chevrolet and Cadillac? Another 6.2 million. And there are plenty of other brands that are making several million cars each year, including Hyundai, Kia, Ford, Honda, Nissan, Suzuki, and BMW. Even the all-electric manufacturer Tesla, which only moved into volume production with the launch of its Model S in 2012, made almost two million cars in 2023 as well. All told, the International Organisation of Motor Vehicle Manufacturers estimates that just over sixty-eight million new cars were built in 2023 alone.

By now, if you've never seen a car production plant before, you may well be wondering how it's possible to build such an incredible number of cars. The first technique that enables output of this scale is the use of what's called a production line. Let's say you want to start building cars and you're going to do it in your garage: you start ordering the parts, and fabricating what you need, and begin building the car in the middle of your workspace. You move around the car, you bring parts to the car, you bolt them on, and work until it's complete. Then a company sees your car, tests it, likes it, and orders ten more. So, you get a larger unit and a few staff, and start working on ten cars. But all this running back and forth to each car, working on them individually, is a slow, laborious process. It also starts annoying your colleagues, because they must constantly move tools, parts, and themselves, from car to car to complete countless different tasks.

To your surprise, you then get an order for 1,000 cars and frantically find yourself reaching for the phone number for a manufacturing company, desperate for some support. Worry not, they tell you, just design, engineer and employ a production line: a series of stages that the cars move through, with staff working on a specific task at each stage, until the car is complete. But you can't roll the car from stage to stage, as it doesn't even have any wheels at the start, so you set up a conveyor system that transports the bare chassis or body, and parts are bolted up, components are installed, and checks are carried out, all as the car gradually moves through your new manufacturing facility. Now, the daunting process of putting a car together from scratch has been broken down into small, controllable steps, each carried out by specifically trained staff at their own stations, and your efficiency and production goes through the roof. The benefits extend beyond just making more cars, too. Because each production step is now more straightforward, which helps maintain consistency and quality, the staff don't need as much training to complete their tasks, keeping your costs down. Further, because you've streamlined your production, you don't need as many people to build the same number of cars, gaining you another thumbs-up from your accountant.

Fifty years later, your car company is putting together almost one million vehicles a year, each consisting of north of 30,000 parts, and its production facilities employ around 80,000 people. Sheet steel is stamped into the body parts your cars need on site, organised, then moved on to dedicated welding lines where the panels are stitched together. Other parts, such as the doors, are then hung from the body, then inspectors check each car and make sure everything's where it's supposed be. The bare bodies are then cleaned, prepared, painted and dried, before more checks are carried out, then assembly begins in earnest: parts are either made on site, or brought in, and as the car rolls down your production line, it starts to take shape: wiring harnesses, engine, subframes, steering, suspension, brakes, they all get hooked in place and bolted up, and then the interior and glass goes in. Some of these assemblies are supplied and installed as complete units, such as the engine, gearbox, front subframe, suspension, steering and brakes, or entire dashboards, making assembly even quicker. And, once that's done, final checks are carried out, the car is tested, then it's ready for dispatch. Every day it is operational, over 2,500 trucks and 120 railway cars arrive at your vast factory, providing the raw materials and parts, from thousands of suppliers, required to build your cars. You review the figures at the end of the day and, yes, almost 230 rail cars and 235 trucks are departing daily from your plant, carrying just over 4,000 new cars. And those aren't finger-in-the-air figures: those are based on the consumption and output of real facilities.

While the production line is one of the key concepts that allows these factories to churn out countless cars, there's one other essential element required to maintain such throughput: automation. Many of the processes involved in putting a car together, especially if you're building lots of them, are exceptionally repetitive, so robots are used to carry out lots of the tasks in a car production factory. A typical car, to put a figure in mind, might contain some 5,000 spot welds alone. To do a single spot weld involves briefly clamping two pieces of metal together using two tongs, like chopsticks, with electrodes at the ends. The electrodes contact a small spot on the metal panels, pass electric current through them, making them hot and melt together, which joins them with a neat weld. It's a quick but precise process, one that must be carried out millions of times each day in a big plant, and robots make swift, reliable and accurate work of it. Robots can also be used to bring stamped parts straight to the moving production line, hold them in place, weld them together, apply sealant, and paint and dry the body shells, allowing the human workforce to focus on the final assembly, which often requires more complex manipulation, flexibility and decision-making.

As you may have guessed, manufacturers didn't instantly start out with vast production lines, stacked with machines, producing thousands of cars at the drop of a hat. No, just like your fictional car company, they started out with smaller, hands-on facilities. Take Carl Benz, as a case in point: the first car he built in any quantity was the Benz Velocipede, a small two-seater car, with four wheels, that was unveiled in 1894. Over seven years, a total of 1,200 were slowly built, making it the first volume-produced car. When the scales really started to shift was in 1901, when Ransom Olds, one of the pioneers of the American automotive industry, took two concepts and revolutionarily employed them in his new car factory: the production line, featuring workstations that the cars would be pushed between, and interchangeable parts, meaning workers didn't need to fiddle with every piece to make it fit. The first mass-produced automobile, the Oldsmobile Model R, started rolling off the lines in the summer of 1901. A total of 425 were built in that year alone, as things got rolling. In 1902, Oldsmobile produced approximately 2,500 Model Rs, also known as Curved Dash runabouts. In 1903? Almost 4,000 cars.

It was Henry Ford who really set the automotive production line in motion, though, quite literally. He founded the Ford Motor Company in 1903, with the aim to develop a straightforward low-cost car for the wider American market. Achieving this goal, however, would require unheard-of levels of efficiency and cost control, and Ford's first cars, such as the Model A and N, were still expensive and slow to build. And even when he launched the now-famous Model T in 1908, a far more reliable and practical car than many contemporaries, it still took some twelve hours to build and cost $825 in basic form. The average annual salary back then, across all industries, was around $550, making the Ford a still-expensive proposition. And while 10,660 Model Ts would roll out of the Ford factory in 1909, the complexity, time and costs of its assembly were still too high.

A dramatic shift was required, so efforts began to tackle inefficiencies in production, in the same way that you faced down the problem in your fictional factory earlier. More machines were brought in, assembly was broken down into separate steps, worker motion was studied and improved, and wastage cut. And, gradually, a series of moving production lines were introduced, emulating the conveyor system that was used in many other industries and advancing the car production line showcased by Olds. By 1913, the time required to build a Model T had dropped to just ninety minutes, permitting savings that allowed Ford to cut the price to $525, and 170,211 were assembled that year. And the improvements kept on coming, as the facilities expanded and production methods advanced: by 1923, Ford was making two million Model Ts a year, accounting for over half of America's car output, and prices dipped below $300 in 1924, about a fifth of what the average American was earning annually by then. The affordable automobile had arrived in earnest and, by the end of production in 1927, Ford had built over fifteen million Model Ts.

6

The Extreme Ends
of the Scale

The automotive world is chock full of variety. Spend five or ten minutes watching the traffic and you'll see countless different types of car and brands, from humble mainstream hatchbacks to expensive lesser-spotted sports cars, and if you dig below the differing surfaces even more diversity will be unveiled: distinctive hardware, interior design, materials, paint finishes, statistics, world-first features, something niche, something bespoke, something rare, and sometimes even something unique. There's so much out there that, if you want to start building your automotive knowledge, or to refresh or develop it, knowing where to start can be difficult. It's like wanting to learn a language and opening a dictionary: you'll just be presented with tens, if not hundreds, of thousands of words, and no distinctive elements that help you start comprehending what's at hand.

When it comes to the automotive landscape, and building a working understanding of the car world, it's best to start at the most basic end of the scale: the utilitarian, mass-produced and hugely popular cars of the twentieth century that helped get people moving in earnest. Sure, some of the slowest, most basic cars around might not seem interesting or notable, but they were the first point of contact with a car for many, or the first car they owned. And, although it's a case of different strokes for different folks, the engineering required to allow them to achieve their goals can prove fascinating. The two you'll probably hear mentioned frequently, Model T aside, or the two that have already popped into your head, are the Volkswagen Beetle and the Citroën 2CV.

The Beetle had its origins in Nazi-era Germany, prior to World War Two, when Adolf Hitler was pushing for developments in Germany's economy and society. By 1933, he'd already started the construction of thousands of kilometres of a new motorway system, called the Reichsautobahn, a network of high-speed, long-distance roads that would boost the economy and cut unemployment while encouraging the new, modern school of motoring. Germany already had some motorways, and the Nazis had previously disagreed with the concept of building more, but Hitler recognised the propaganda potential of mobility and changed tack completely. But these new roads and this popularity push would be for nothing, however, if no one could afford the cars that were currently available. Engineer Ferdinand Porsche, founder of what was then the engineering consultancy firm Porsche, was tasked in 1934 with creating a practical, affordable and reliable car that would serve the new German motorists well. The result was the KdF-Wagen, a two-door car with a basic air-cooled engine out back, driving the rear wheels, a plain interior that offered room for four, and a luggage storage compartment in the nose. It had a simple chassis, to which everything mechanical was bolted, similar in concept to the modern 'skate-board' platform used in electric cars, and the body was dropped on top. And, in 1937, a company was founded to build this new car: Gesellschaft zur Vorbereitung des Deutschen Volkswagens mbH. A company to prepare the way for the German people's car, 'Volk' meaning people, 'wagen' meaning car.

The war, unsurprisingly, squarely torpedoed any chance of getting the new KdF-Wagen into production but, in June 1945, the British Military Government took over the factory and started series production. It needed vehicles, so it and the country could continue to operate, and the Germans needed jobs. Before long, thousands of what were now called Type 1s were being cranked out each month, and the affordable car started making its way to drivers and fulfilling its intended role. Its inexpensive, simple, and economical nature gained it footholds in countless other countries, including America, where it became a counter-culture symbol during the 1960s. In the end, what was a product of Nazi-era Germany, freed from its fascist shackles and stoked into life by the British, evolved into an automotive icon that ended up being adored by millions around the world, first nicknamed and later officially titled 'Beetle'. In the end, when production stopped in July 2003, 21.5 million had been built.

The famous Citroën 2CV was born out of a similar mobility-boosting concept to the Beetle: the need for uncomplicated, capable cars that could be afforded by those who could benefit from them, such as farmers and people living in remote areas. What Citroën came up with was a masterclass in a product being fit for purpose. The late prototype of 1939, called the TPV, Toute Petite Voiture, Very Small Car, weighed around 400kg, could carry four people and 50kg of luggage, reach 31mph (50km/h) fully loaded, and could roam its way across rough terrain without issue. It was elegantly engineered, with innovative suspension, yet affordable and inexpensive to run. Like the Beetle, it employed an air-cooled engine, albeit with two cylinders, not four, it was easy to repair and maintain, and its body could also be lifted off its chassis. The TPV also had some big benefits over Volkswagen's Type 1, including four doors, removeable seats, and a canvas roof, all of which effectively allowed it to function as a van. But, also like the Volkswagen, the reveal and production of what would be called the 2CV at launch would be delayed by the war. It wasn't until October 1948 that the Citroën 2CV took centre stage, but the inexpensive, eminently practical and straightforward car proved an immediate success. What became charmingly dubbed the 'tin snail', due to its distinctive looks, would also become a hugely popular car around the world, with almost four million being built by the end of production in 1990. If you include later versions, such as the Dyane and Ami 6, that number swells to almost nine million.

The Beetle and 2CV weren't the only mass-volume affordable cars of their era, though, with others including the oft-overlooked Renault 4, the famous Morris and Austin Mini, the Fiat 500, Morris Minor, and later the Trabant 601. If you're looking to build your automotive knowledge base, all are worth looking into. But companies such as Citroën weren't just about basic and affordable motoring, though: some of its cars swung right to the other end of the scale, being luxurious, refined, advanced, and complicated. Among the most famous cars in the world, both for technology and beauty, for example, is the Citroën DS. The streamlined four-door executive saloon, introduced on 6 October 1955, must have looked like a UFO to the assembled crowd at the Paris Motor Show. Its engineering was just as remarkable: it had self-levelling oleo-pneumatic suspension, with adjustable ground clearance, a hydraulic pump that provided power to its suspension, steering, brakes, and semi-automatic gearbox, inboard front disc brakes, and a host of safety features, including a padded dash, collapsible single-spoke steering wheel, and thin windscreen pillars that were designed to improve the driver's visibility. Later versions would also receive features such as steerable headlights, which turned with the steering to offer a better view of what was ahead.

If it's outright luxury that piques your curiosity, it's brands such as Bentley, Rolls-Royce, and Mercedes-Benz, that should otherwise be at the top of your to-study list. Take the Mercedes-Benz 600 of 1963, also known as the Grosser Mercedes, for example: this vast 2.6-tonne luxury saloon featured a mechanically fuel-injected 6.3-litre V8, a four-speed automatic transmission, and could accelerate from 0–62mph (0–100km/h) in 10 seconds. It also had adjustable self-levelling air suspension, four-wheel disc brakes, power steering, central locking, dual-zone heating and air conditioning, as well as hydraulically powered seat adjustment, doors, boot lid, windows, and optional sunroof, and each car was finished to incredibly high standards. They were ferociously elaborate and, even today, many regard them as one of the most complicated cars ever built. They unsurprisingly suffered mechanical issues right out of the gate and buying, maintaining and using one required seriously deep pockets. Even so, 2,677 were built over a seventeen-year run, with notable owners including Elvis Presley,

Jack Nicholson, David Bowie, and, uh, Mao Zedong, Saddam Hussein, and Robert Mugabe. Yes, the vast, imperious, significantly-richer-than-you Grosser was unsurprisingly popular among dictators.

If you're more interested in speed, acceleration, and exotic construction and design, you need to dig into the land of supercars and range-topping hypercars. Companies such as Ferrari, Porsche, Lamborghini, Aston Martin, Lotus, Koenigsegg, Rimac Automobil, Pagani, McLaren, Maserati, GMA, Czinger, and Pininfarina, among others, build the quickest and fastest cars around, the ones that thrill- and prestige-seeking buyers will gladly pay hundreds of thousands, if not millions, for. Some of the most notable in history, and ones which are referenced often, include the twin-turbocharged Ferrari F40, the elegant Lamborghini Miura and angular Countach, the driver-focused McLaren F1, and the all-wheel-drive Porsche 959. But their performance pales in comparison to their modern counterparts. For example, a 1987 Ferrari F40 packs a still-impressive 471bhp (351.5kW) and, if you did away with every modicum of mechanical sympathy and launched it perfectly, it'd accelerate from 0–62mph (0–100km/h) in 4.1 seconds. Ferrari's modern SF90 Stradale, on the other hand, produces 986bhp (735kW) and goes from 0–62mph, with zero fuss, in 2.5 seconds.

To put the capability of modern performance cars in further perspective, Koenigsegg's Jesko Absolut, revealed in 2020, can accelerate all the way to 249mph (400km/h) and back to zero again in just 27.83 seconds. If you lined up on a track against it in the most powerful version of the Citroën 2CV ever made, and the flag dropped, the Koenigsegg would get to 249mph, then brake to a complete stop, several seconds before you'd even hit 60. There are still plenty of slow cars around, though, as manufacturers continue to make cars that are focused on being fit for purpose. A town runabout doesn't need several hundred horsepower, after all, or the heavy, expensive and complicated brakes, suspension, steering, chassis, cooling systems, and the rest, that high-power cars require. They don't even need 100bhp, just to get someone around, as cars such as the Beetle and 2CV ably demonstrated. And, as you've already read, you don't need lots of power to have fun in a car.

How about the most beautiful car in the world? Well, that's a distinct matter of opinion. A lot of people will tell you it's the Jaguar E-Type, a sleek two-door sports car that made its debut in 1961. Others might cite the earlier XK120, made between 1948 and 1954. There are a lot of Italians that could justifiably take the number one spot, including the Lamborghini Miura and Ferrari 250 GT Berlinetta

Lusso. One or two might even stick their hand up for the early second-generation Chevrolet Corvette, with its distinctive split rear screen, or the mechanically direct-injected Mercedes-Benz 300 SL 'Gullwing' of 1954, which many regard as the world's first supercar. The art deco fans, on the other hand, would probably tip towards something such as the Bugatti Type 57, perhaps the SC Atlantic Coupé, or other similarly dashing coupés and roadsters from the 1930s. There are contenders from more recent times, too, including the V12-powered McLaren F1 and the record-setting Bugatti Chiron. The ugliest cars? Well, again, like many things, it's a highly subjective topic. If you're about to point your browser at your favourite search engine, your starters for ten should include the Pontiac Aztek, the SsangYong Rodius, the Mitsuoka Orochi, the Allard P2 Safari, the Tesla Cybertruck, and the Fiat Multipla. But even these have their fans, their appeal, and their place. A true automotive enthusiast, in any instance, will always, at the very least, understand and appreciate that others can like something they may not.

7

Diesel, Petrol, and Electric Power

Let's talk about petrol and diesel for a moment. The former, which is also often called gasoline depending on where you are in the world, is a flammable liquid produced by refining crude oil, or petroleum, which we extract from the ground. Yes, the stuff that's formed from ancient organic matter being compressed and heated for millions of years. Most refineries, before the advent of the car, simply did away with petrol that was produced during refining, as it didn't seem to have any good application. It was a similar story for diesel, another unwanted byproduct from the refineries which were extracting kerosene for lamps from the petroleum, and it was originally just referred to as a type of fuel oil or distillate. That would all change in 1897, when an inventor and engineer called Rudolf Diesel went to start a prototype of a new engine for the first time.

In a conventional spark-ignition petrol engine, the compressed fuel-air mixture is ignited using a spark, and you know what happens after that: whoomph. But this has its limitations, partly because if you compress the volatile mixture too much, it'll get hot enough to cause auto-ignition, where it spontaneously ignites. What you tend to get then is an uncontrolled and erratic bang, instead of a controlled burn, and if it happens at the wrong point, it can cause significant damage to the engine. This limits the amount of compression you can use, which in turn limits the efficiency of the engine. You can't squeeze the mixture too much, so it can't expand as much, meaning it can't do as much work as possible on your engine's piston. And the early engines, like those built by Benz, and coal-powered steam engines, were inefficient. What Diesel wanted to do was to create an engine that was more economical, cleaner, and could run on a variety of accessible fuels, giving new and smaller businesses a way to mechanise their operations.

The engine that Diesel subsequently developed, the eventual fuel and design of which would end up bearing his name, differed from spark-ignition engines in two important ways: it used less volatile fuel, and the fuel wasn't added until the air in the cylinder had been compressed. Instead, the fuel was forced or injected directly into the cylinder once the air had already been squeezed and heated, at which point it would ignite of its own accord. This and the more stable, slower-burning characteristics of the fuel allowed the compression ratio to be higher, making the engine more efficient, while also ensuring the combustion process of the fuel started at the right point. This injection and auto-ignition is admittedly still more violent than the combustion cycle in a petrol engine, and the sudden spikes in cylinder pressure create the characteristic diesel clatter that puts some drivers off diesels entirely. The engines are also typically heavier because of the diesel combustion process, because they have to withstand the forces resulting from the high compression and cylinder pressures. And while these compression-ignition engines have their disadvantages, their increased efficiency and increased torque often makes them a desirable choice. Diesel also contains more energy than petrol, so you can use less to achieve the same amount of work.

Like any engine, though, there are snags. Regardless of fuel type, and pollution and inefficiencies aside, any substantial automotive engine is a complicated affair. Hundreds of parts, some spinning at thousands of revolutions per minute, or moving at myriad metres per second, incessantly. It takes a lot of effort to keep one intact and operational, especially for any useful amount of time. There's lubrication, where oil is pumped and splashed around, to stop friction from ruining surfaces and causing parts to seize. Then there's cooling, where air or coolant is circulated around the engine to avoid damage and inefficiencies. No matter how well it's designed, how well you get air in and out, how precisely you meter the fuel and control the combustion process, there are limits to how efficient engines can be. The figures will vary significantly depending on where and what you look at but, as a finger-in-the-air example, of the fuel supplied to an engine, you'll be doing well if more than thirty-five per cent of it is converted to useful work. The rest will be waste: heat, noise, and a mass of exhaust gases and particulates, all things that end up causing problems down the line.

There is a more efficient and less-complicated alternative: electric power. You store some energy in a battery, generated by a more efficient power station or renewable energy source, and use it to drive an electric motor. The number of significant moving parts in that motor? Usually just one: the rotor. And the efficiency of an electric motor? In some instances, over ninety per cent of the power supplied to it will be used to propel the car forward, knocking the spots off an internal combustion engine. Then there are the other obvious benefits: they're quiet, they're smooth, and there are zero local emissions. Electric cars can also be fun, offering emission-free driving experiences, ones that liberate you from the need to refill with expensive fuel, to endure costly servicing, and to tackle the numerous mechanical problems a conventional car can develop. Sure, a lot of the run-of-the-mill ones might not be that interesting to drive, but then the same can be said for plenty of conventional petrol and diesel cars.

Electric cars are not magical machines, however, and they do have their foibles. These include oft-prohibitive purchase costs, charging issues, and range limitations, all of which can make them a poor or simply inaccessible choice for many drivers. Yes, concerns about power generation to charge the cars, rare earth and mineral extraction to build the motors and batteries, and matters surrounding the speed at which they are being introduced and legislation phasing out internal combustion, are valid concerns, but often overplayed. As a case in point, harmful lifecycle carbon dioxide emissions for an electric car, including building it, charging it, and ultimately disposing of its battery, are on average a third of that of an equivalent petrol car. Plus, greener ways to build batteries, and charge them, are being developed and employed. It is not static: like the advancement of the engine, and the infrastructure to support it, the technology will quickly change and improve with more focus on it to realise its advantages.

What's interesting about electric cars and their situation today is that much of it is just a little bit of history repeating itself. Like some other technologies, electric cars have been around a lot longer than many might realise: the first electric cars made their debut in the 1830s, with non-rechargeable batteries, and in 1884 a rechargeable electric car was built by British inventor Thomas Parker. These clean, green machines weren't niche, either: in America, in 1900, the census revealed that, of the 4,192 cars produced that year, 1,575 were electric. Yes, almost 40 per cent of America's vehicle output was electric, and electric vehicles accounted for around a third of all vehicles on the road in the country at the time. They were understandably popular because they were cleaner, quieter and less complicated than the early and finickety internal-combustion-powered alternatives. You didn't have to deal with hand-cranking an engine to start it, potentially breaking your wrist in the process, you didn't have to deal with water, oil, fuel, or any other of these hassles, and you arrived at your destination free from the odour of combustion and crankcase blow-by.

However, the range of early electric cars was particularly restrictive, charging facilities were few and far between, and they could be expensive. Sounds familiar, right? And, as the cheaper internal combustion engine got more refined, reliable, and capable, and fuel became more available and affordable, electric cars got left in the dust. If things had been a little different, if development and acceptance of electric cars had continued in the early days, the automotive landscape of the twenty-first century might look very different. That's one of the things that's worth remembering about today's electric cars: they haven't benefitted from the hundred-plus years of refinement and advancement that petrol and diesel cars have, yet they've already made terrific strides in capability in a comparatively short amount of time.

As with the Model T, the increasing demand for, and manufacturing of, electric cars is also helping to bring prices down, and ever-advancing battery technology and charging solutions further help make zero-emissions motoring more realistic and viable. Here's a good example of just that: when the first-generation Nissan Leaf, the first mass-market modern electric car, went on sale in 2010, it cost £28,990, excluding government grants, and had a claimed range of 100 miles (161 kilometres). In 2025, just fifteen years later, it'll cost you half that for something with a similar range, while spending around £30,000 can get you into a car with a claimed range of north of 300 miles (483 kilometres). The take-up of electric vehicles, prompted by falling costs, increasing awareness, and a range of options, is vividly evident when you look at the change in market splits. In Europe, for instance, electric cars accounted for just one per cent of total new car registrations in 2018. Just six years later, that figure had grown to 13.6 per cent. In 2024, in Norway where the adoption of electric cars has long been promoted and incentivised, almost ninety per cent of new cars sold were electric.

Echoing the surge in popularity of bicycling, the rising sales and availability of electric cars has led to a burgeoning used-car market. Which, in turn, has resulted in numerous specialist companies being established to help support and maintain older electric cars, further widening the choice for car buyers. That isn't to say that electric power suits everyone under the sun, and it may never fit the bill for certain markets or situations, at least not for a long while. But, for many, it may rapidly become the desired choice, the go-to option. And to disregard it, at least unthinkingly, could cause you to miss out on what you might find to be a gratifying form of motoring. The absence of vibration, noise, and emissions, the surge of acceleration, along with facets such as the ability to jump in and drive without having to worry about warming up your engine, grant electric motoring its own distinct charms. At the very least, it's worth trying a decent electric car. Even if it's not to your liking, you'll probably see why people make the leap, and continue to do so in increasing numbers. The electric car revolution is coming, some sources may tell you. The truth of it is, in some countries, it's already happened.

When Manufacturers
Get It Wrong

My body wasn't trying to eject itself through the windscreen, which was the first sign that something was wrong with the Suzuki. The second was the light metallic snapping noise that the brake pedal had made when I stomped on it. I looked at my colleague in the passenger seat, *Autocar*'s road test editor Matt Saunders, and his expression was as baffled and as disbelieving as mine. I glanced down into the footwell and, lo and behold, the brake pedal was on the floor, along with my foot. The latter was pleasingly still attached to my leg, but the former appeared to no longer be attached to anything. I thought about the corner of the test track that was approaching at some 80mph (129km/h) but decided that was a problem for future me, and that I had enough time to investigate this issue for a moment. Tentatively, I pulled the pedal upwards with my toe, then pushed it down again, but it just flapped around like a leaf in the breeze. It was 30 January 2015, and the Celerio, Suzuki's then-new small hatchback, had suffered complete brake failure.

Despite the resources of the company, the efforts of countless engineers, and all the tests and trials that a car goes through before it reaches the road, something was evidently wrong with the brake pedal retraction system. In a serious collision, when forces above that of emergency braking were applied to the pedal, a component in the assembly was designed to fail, allowing the pedal to drop away and stop it from hitting the driver's legs or feet and causing injury. However, as mashing my boot down on the pedal during our regular brake-distance tests had demonstrated, the system in our test car was too sensitive: if you tried to do an emergency braking manoeuvre, the pedal would prematurely detach, leaving you with no brakes. Fortunately, having the luxury of an empty test track, I simply used the gears and the handbrake to bring the car to a controlled stop. What did make my hair stand on end slightly, however, was the fact that my colleagues had been driving the car on the road beforehand. Luck, evidently, had smiled on us and those around us.

Suzuki, as you would expect and hope, acted immediately. The car that had experienced the failure was collected for inspection straight away and exchanged for another, so our testing could continue, because the issue was hopefully a one-off. Perhaps, as sometimes happens, a parts supplier had delivered a faulty batch of components. In these instances, careful checking of part and production numbers would allow the company to identify the cars that had potentially been fitted with faulty pieces. However, ten minutes later, we were sailing down the mile straight at the Millbrook

Proving Ground with no brakes again. The problem was potentially more widespread, and Suzuki triggered what is known as a recall: a process where, once a safety issue has been identified by a manufacturer or external agency, owners are contacted, a solution is developed, and dealers carry out the required repair. The car was just coming on to the UK market, so there weren't many on the road, but Suzuki contacted and informed the owners, halted its dealer activities and promotions, and its engineers went to work solving the problem. It was subsequently discovered that, although extensively tested originally, some of the production parts were bending when they shouldn't, causing a failure. Just ten days later, chief engineer Shigeki Suzuki invited us back to Millbrook to demonstrate the revised pedal assembly. It featured stronger components, which could better resist bending during unexpectedly high loads, preventing premature detachment while still otherwise doing their job. The car then flew through our tests, stopping safely and controllably, several times over: the problem was fixed. And, happily, no harm was done.

While a recall might sound like an exceptional situation, they are commonplace affairs, and increasingly so as cars continue to get more complicated. In America, for example, 1,000 recalls were issued by the National Highway Traffic Safety Administration (NHTSA) during 2023, accounting for a whopping total of almost thirty-five million vehicles. In the UK, the Driver and Vehicle Standards Agency (DVSA) also issued over 800 recalls in the same year, which accounted for some 3.8 million vehicles. Sometimes, a recall will be triggered for a relatively insignificant issue: Volkswagen, for example, once had to recall 3.7 million cars in 1972 due to a defect with the wiper arms that could cause them to come loose. Sometimes, the recall is due to something completely unexpected. Mazda, not once, but twice, had to recall tens of thousands of cars due to a breed of spider that was attracted to the smell of petrol and built webs inside its cars' emissions control systems, potentially causing problems with the fuel tanks. Not all recalls are large, either. Supercar manufacturer Koenigsegg once recalled just a single car, an Agera built in 2012, due to a fault with its tyre pressure monitoring system.

Sadly, some recalls are the result of far more significant issues resulting in deaths, which are not all handled well. The General Motors ignition switch recall is a prime example of such a situation, spurred on by cost-cutting measures, financial pressures, poor communication, and an improper corporate culture. At least as early as 2001, GM engineers noted that, in some cases, it was too easy to turn a key in one of the ignition switches it was using. This meant, if it was bumped or had something hanging from it, the engine could shut down, or the car could switch off completely, leaving the driver without steering assistance and potentially disabling the airbags. The loss of control, and the loss of power, could easily cause an accident, and one where the safety systems might not operate properly. The problem was reported again in 2003, and again in 2004, but nothing was done. More reports followed, with no response, then a death in 2005 was linked to the ignition switch problem. Yet the issue persisted and no recall was issued. The National Highway Traffic Safety Administration, a government regulatory body that could instigate a recall, didn't have enough information to act on, and GM's lack of communication compounded the problem. Eventually, as external pressure mounted, as more incidents and deaths were related to the fault, GM eventually started a recall in early 2014. Initially, the company recalled 619,122 cars. Two weeks later, an additional 748,024 vehicles were added to the recall. Then another 823,788. The number kept growing. By the end of the year, GM had recalled almost 30.1 million vehicles globally, at least half of which were directly related to ignition switch or key issues. But, for many, it was too late: in the end, the defect was linked to 124 deaths and 275 injuries. GM would pay a $900 million (£580 million) fine to the government to avoid a criminal investigation and end up paying $575 million (£370 million) in settlements, but many felt the financial cost and punishments were not significant enough. Particularly when, in 2014 alone, the company still made a $2.8 billion (£1.9 billion) profit. The costs would stack up, though, with GM eventually paying out north of $4 billion (£2.6 billion) in total. And the cost to fix the issue, as identified and calculated by the company itself in 2005? Just fifty-seven cents, or less than forty pence at the time.

There are countless other infamous recalls, including the Firestone-Ford controversy, which prompted a major recall in 2000, Toyota's unintended acceleration defect of 2009–2011, and the wide-ranging Takata airbag recall, which started in earnest in 2013. There have also been a few rare situations where the manufacturer has been found to not be the root of the problem. Sometimes, it's PEBWAS: Problem Exists Between Wheel And Seat. Those stories about cars careening out of control, as a case in point, reputedly accelerating without driver input? Well, many end up being the result of simple driver error, where the accelerator has been mistaken for the brake pedal, and the ensuing panic means the driver doesn't recognise and correct their error. See also: PEBKAC, Problem Exists Between Keyboard And Chair, which hopefully isn't what you're thinking about me as you make your way through this book.

When Manufacturers Get It Wrong

But it's not just recalls, or panicking drivers, that cause disasters for manufacturers. The road to success, as the saying goes, is littered with failures, and sometimes companies make errors of judgement that end up ruining their new car, or the company's annual reports. Take the first-generation Smart ForTwo, which was an innovative compact city car launched by Daimler-Benz, later Mercedes-Benz. Too much money was spent on developing it, and building its brand from scratch, and the demand and sales potential wasn't there to support such a hefty investment. Bernstein Research, a financial and investment research company, later calculated that the ForTwo ended up costing the company around £2.9 billion between 1997 and 2006, about a year's profit. Or how about the Fiat Stilo of 2001? It was designed to take on the fabled Volkswagen Golf, but it wasn't competitive, didn't sell well, and cost the company £1.8 billion. The CEO at the time, Roberto Testore, promptly resigned, and avoiding complete collapse required a dramatic reorganisation and a tranche of new models. As luck would have it, one of those new cars turned out to be a remarkable reinvention of the original Fiat 500, which went on to be so popular that it pulled the brand back from the brink. But even remarkable and terrifically expensive cars, such as the Bugatti Veyron, were often expensive endeavours: Bernstein estimated that, after development costs, each of the circa £1 million Veyrons built was costing its parent Volkswagen Group £3.9 million.

The list of cars that have not gone down well in history is extensive, ranging from the inoffensive but inadequate, such as the Hillman Imp, to the intolerable and inexcusable, such as the Chrysler TC by Maserati. But, in the past, some companies have made errors of judgement so significant that they've changed the entire course of the automotive market in which they operate. As America struggled with the 1973 oil crisis, and government fuel economy targets went through the roof, customers sought more frugal motoring options. Subsequently, General Motors had the bright idea to introduce an economical diesel engine into its lineup. It wanted to pinch pennies, though, so it didn't develop all-new engines and tooling. No, instead, it took an existing petrol engine and redesigned it to use diesel, then unceremoniously dumped it into an array of Oldsmobiles, Buicks, Cadillacs, Chevrolets, GMCs, and Pontiacs. Unsurprisingly, given the engineering shortcuts and cost-cutting elsewhere, the engines were noisy, dirty, underpowered, and unreliable. To such a degree, in fact, that every single one of the nine cars submitted to the Air Resources Board in California, for emissions testing in 1979, suffered from engine problems. And customers found it just as bad: lawsuit after lawsuit followed, because of problems with the cars, causing some states to establish the first 'lemon laws', legal actions which could force manufacturers to buy back cars that didn't meet expected standards. GM would persist, though, and had fixed many of the problems with the diesel engine by the end of its production life in 1985, but the initial damage tanked the appeal of diesel as a fuel for passenger cars in America. It would take almost thirty years before GM dared employ diesel in its cars again, in the Chevrolet Cruze in 2013. But a year later, the infamous emissions scandal, known as Dieselgate, began. Volkswagen, among other manufacturers, was caught cheating in emissions tests, kicking off a whole ream of ugly lawsuits and recalls, scuppering the diesel automotive market for years to come.

As many of these failures ably demonstrate, and as you can no doubt imagine, cars are difficult to design, difficult to build, and events often difficult to predict once they get out into the real world. What you want to pay particular attention to, at the end of the day, is how the brand in question recovers from its missteps. Are the conversations about the stumbling blocks open, transparent, and the resolutions prompt? Then you may deem the organisation sound, its responses demonstrating the right kind of approach to its customers. You are potentially going to give them lots of money, so you want them to show a vested interest in your safety and satisfaction. But when a company hand-waves problems away, ignores its issues and consumers, or tries to brush over its tracks, well, you may rightly consider it undesirable to have one of its products on your drive. In time, if enough buyers vote with their feet, the ensuing pressure can result in a change for the better.

9

A Potted Guide to
Buying a Car

People often ask me for help when it comes to buying a car and, in a stereotypically polite and British fashion, I'm ever ready to help. I'll sit down with them, talk about what they need, walk them through the options, work with them to identify a few cars worth looking at, and even drive out with them to inspect a car they're interested in. Naturally, after they've gone and thought about it, they'll then disregard all the advice they've received and buy something completely different. A week later, inevitably, the complaints start. It's unreliable. It's inefficient. It's not that good to drive. It's not practical. Its range isn't long enough. And so on. But I can understand this turn of events, especially if you have little to no experience of buying a car: as well as potentially being a daunting, complicated and stressful process, it's also easy to get carried away and end up with something that doesn't fit your bill.

Even if you've been through the car-buying loop several times, it's still easy to make a similarly misjudged purchase. You know, the car in the classifieds, or the one on the forecourt, that catches your attention for a moment. Perhaps it's the colour, the specification, the way it looks, or just its price. You know it's not quite right for you, but there's something about it. The next thing you know, the seller is talking you through what you could do to make a deal today, and you start feeling like a boulder tumbling down the hill at an ever-increasing rate of knots. Then you're testing it, skipping over your checks and inspections, and you start to discuss money. Then you realise how much time you've spent there, and you start feeling like you should just buy it and go home, so you gloss over the issues with a smile and shake hands. Now, you're driving away in it and, oh no, it's still outside when you open your curtains the next morning and, yes, it's your problem now. You try to convince yourself that it'll be okay, but the doubts creep in, spoiling the experience, and it doesn't work out. Eventually, you're forced to bail out of it, sell it for a loss, and have to go through the whole process again. So, please, if you can, try to avoid this. Just step back and politely decline if you're not sure or comfortable. You're not obliged to buy the car, or any car, and if you walk away then you could save yourself from lots of problems. When it's right, you'll know it; when it's not, as is the case with much in life, saying no can ultimately put you in a better and happier position.

Fortunately, when it comes to buying a car, there are some easy to ways to help ensure that you're happy with your choice. The first is to make a list of what you need from the car, and what you want. Getting some specifics and preferences down on paper, ranging from seat count to fuel type, and everything in between, can help you quickly cut the number of cars that you have to look at. If you don't do this, and you start browsing, you can easily get overwhelmed by all manner of options that, eventually, will turn out to be dead ends for you. Before you know it, you've sunk countless hours, and possibly a chunk of money, into investigating cars that won't ever be suitable. There are some less obvious practical points you might want to consider, too, aside from ones such as interior space. The size of many a car, as a case in point, can make it a chore to live with, so keep an eye on its footprint. Wincing every time you see oncoming traffic on a narrow road, or battling with every parking space, can get real boring, real quick, if you don't need a large car. And if you're tempted by an electric car? Then, aside from range, you'll want to think about facets such as whether you can get a charger installed at home, where you will charge it otherwise, and so on.

The big and obvious consideration, be it buying new or used, is to fully account for the financial side of things. And not just the purchase price, or the potential costs of a finance deal: you need to think about running costs, such as insurance, fuel, and servicing. Additionally, and crucially, depreciation. That major factor, the change in a car's value due to its age, mileage, and condition, can be a significant cost that's often overlooked. For example, if you buy a new car outright and it loses a third of its value after a year and 10,000 miles (16,000 kilometres), that's a big stack of money that's effectively just disappeared. If you're trying to keep costs in check, and having an all-new car just isn't important, it's better to just buy something a little older and dodge the worst of the depreciation. You need to keep costs in mind when you're viewing used cars, too. For example, you might find a car that meets your requirements, but it needs some work to bring it up to scratch. Those costs might snowball quickly and, eventually, it could end up being less expensive and stressful to just buy a better car in the first place. Likewise, if you're considering an older car, or one that's been modified or restored, the pitfalls are bigger, and the potential costs are higher. Consequently, you need to carefully research, assess, and understand what you're looking at before you jump in with both feet.

Your car-buying experience, and what you'll need to do, will however vary depending on what you're buying and from where. Sourcing a new car from a dealer? Then you'll have a lot of protection, in legal terms and with regards to warranties and getting problems sorted, and you might be able to strike an advantageous deal. Buying privately? Then you'll need to really do your checks, and make sure you've got your eyes open, as there's often little recourse should something go wrong. Auctions? Being frank, you're best off bidding with money you can afford to lose. The old adage is that cars are often at an auction for a reason and, while not always the case, it's easy to end up paying over the odds for a car once the bids start flying. More to the point, that car might subsequently not even work properly, and the undetected issues could cost thousands to fix. In some cases, buying privately or from a dealer could end up costing you less, one way or another, and you'd also be able to inspect and test the car before buying. But if it's a rare car, or something unusual, I can understand you wanting to stick your hand up and bid. Just scrutinise it as best as possible beforehand, make sure you understand the auction process and costs, and be aware of what you might end up with, then set your limit and stick to it.

One other titbit worth remembering is that you're buying the car, not its surroundings. It's all well and good that the car you've seen is being sold from an upmarket dealership or house with an appealing gravel drive, but don't let that distract you from what you're interested in: just because the environment is superb doesn't mean the car is. And whatever you do, don't rush. You want to carefully check the exterior, the interior, the underside, the function of every button and switch, to listen to everything, to test it properly, to read its paperwork, to carry out the proper background checks, and to fully satisfy yourself about the car's condition. Advantageously, there are some tells that a car has been well looked after, aside from running and looking the part, and having a substantial history file. The readily apparent one is that the car is clean and tidy throughout, which indicates either a modicum of care or a desire to present the car properly, both of which are desirable. Other good signs include matching high-quality tyres, invoices that show that quality parts and fluids have been used to maintain it, and evidence that small problems have been fixed promptly. In some cases, what look like minor issues can be a pain to sort and, left unattended, they can sometimes turn into big problems, so tread carefully if you sense trouble. The seller, too, can provide a steer as to the quality of the car; if they're keen to show you around, to explain what's been done, or what might need doing, and they readily answer your questions, then these are all good signs.

For many of us, buying a car is one of the biggest investments we will make. But it's just the start of the process and, once you've done all the legwork and bought your ideal car, the ownership phase begins. It need not be taxing, though: just take a moment to familiarise yourself with the car, to read its manual, and to understand its servicing needs and requirements.

Check the essentials regularly, keep it legal, keep it roadworthy, and maintain it, and you should be able to enjoy many miles without fuss. And, once you've got to grips with it, you can start exploring car ownership in earnest. Maybe you'd like to do more than just commute. Maybe you'd like to explore your country, or maybe you'd like to get involved with the world of cars, be it through modifications and personalisation, or car clubs, motorsport events, racing, shows, or local gatherings. There's a lot you can do with a car, aside from just travelling, and there are countless clubs and communities out there to engage with and contribute to.

Don't feel locked into one car or experience, though. If you've the need or inkling to change, if you want to sample something else, and you have the resources, go for it. You can sell your car, buy another, learn something new, and move on to the next. You never know where it might take you, or what or who it might introduce you to. But just be mindful of the costs involved, both financially and mentally, and do what's best for you. And when it finally comes to selling your car, make it as straightforward as possible for yourself and the buyer: write an honest description, list all the car's faults, take lots of pictures, and always be up front about the car's condition and what it needs. It'll save you from hassle, and it'll save the buyer from a wasted trip. But all told, this isn't an exhaustive nose-to-tail guide to car buying and ownership, as there's only so much room here, and there's so much out there, from country to country and from car to car. What it should do, though, is give you a good starting point. And, if you do have a specific question, a quick check online, or a glance at a book or magazine, or a phone call to a specialist or club, should point you in the right direction. Just remember one thing: if you don't like the car you're looking at, if you have doubts, don't buy it. There will be others, and the right one for you might be just around the corner.

10

Autonomy and the
Future of the Car

It is the end of your working day. Back when you started this job, you used to dread the commute home. Now, though, it's something you look forward to. You glance down at your phone and a small icon in the top-right corner of the screen changes from grey to green: your car is outside. It has driven from a dedicated parking complex, a few miles away, to your office, and it's now waiting outside in a designated spot. You make your way out to it, take a seat, press the green button on the screen to start your trip home, cross your arms, and drift off. The journey through the outskirts of the city is smooth and peaceful. The car knows what's going on: it has a Vehicle-to-Vehicle communication system, allowing it to get information on other vehicles' speeds, directions, and locations, allowing it to optimise its route and avoid incidents. As you nap, the car makes its way on to the motorway, tucking in closely behind a convoy of already-rolling cars, the air being pushed out of the way by the cars in front reducing drag and boosting your car's efficiency. Your car knows when the others are slowing, accelerating, or changing lane, so perfect pace and position is maintained. Forty minutes later, it swoops through a junction, into the countryside, eventually rolling quietly to a halt outside your home. It lets you get out, then parks inside your garage. The charger on the wall automatically deploys a robotic arm, plugging the car in, waiting for the electricity rates to fall before charging starts, and the garage door shuts. You slide your access key over the pad by your front door and think, fleetingly, that level five automation is an absolute godsend. It's not just cruise control: the car can handle anything that is thrown at it. It is completely autonomous, requiring no human input whatsoever.

It might sound unfeasible but, even today, the technology to provide such a capability is already being developed and trialled. We already have publicly available cars with driver support features that allow them, with human supervision, to steer, brake, and manoeuvre for the driver, cutting down on the input required to get to your destination. And, in specific areas of certain countries, you can even access fully autonomous taxis that, while they can't yet operate in all conditions, will otherwise take you from A to B with no driver input at all. And the advantages offered by autonomous vehicles, from increased safety and efficiency, through to reductions in congestion, are among the reasons that manufacturers are pushing ahead with the technology. Customer demand, too, is another driving force: the car is a convenience, a car that drives itself doubly so. Many of us enjoy driving, true, but none of us enjoy the process when we're just stuck in static traffic. Automation might also permit you to do away with your car entirely: if you can reliably access an autonomous taxi or car-sharing service, at the press of a button, you might not need to own a car. Commercial transport also stands to benefit, with computer-controlled trucks, buses, and other goods vehicles, operating more efficiently and safely than their human-operated counterparts. For example, imagine a trucking operation where the trucks automatically take the most economical routes, drive in the most efficient fashion, and can operate day and night, non-stop, boosting the throughput and productivity of the company, all while cutting costs.

There are still hurdles to cross, however, including advancing the hardware and software to the point where it is reliable and effective. As you can imagine, autonomy requires a staggering degree of computing power: the car's system must process streams of data from radars, light detection and ranging (LiDAR) systems, cameras, ultrasonic sensors, traffic data, weather information, and high-resolution mapping, to enable them to view and interpret what's happening around them, enabling safe automation of the vehicle. Not all automation is the same, though. Driving automation is defined by SAE International, formerly the Society of Automotive Engineers, using one of six levels, ranging from zero to five. Levels zero to two encompass driver support features, such as adaptive cruise control and lane centring. As their name suggests, these systems provide support, not automation, and cannot handle all conditions. Consequently, the driver must still pay attention, watch the road, and be ready to intervene and assume full control. SAE Level three marks the start of automated driving, where the driver becomes a passenger when the systems are operating. At level three, however, the driver may still be called back into the loop. At level four, the vehicle can operate of its own accord within a certain carefully mapped and controlled area. Autonomous taxis are a good example of level four autonomy. And level five, as you may have guessed, is the full driver-optional system. A car that can go anywhere, drive in any conditions, and handle every aspect of its trip, without any human interaction.

The technological challenges are only a facet of the issues surrounding wider-scale autonomy. Autonomous vehicles might interact with each other in a controlled and predictable way, but what happens when an errant human, a PEBWAS, gets thrown into the mix? Will autonomy cause incidents on the road to fall to the point where the remainder are a result of human error, prompting calls to remove the human element from the equation? And how will these vehicles interact with cyclists? Should we tear up the existing infrastructure and redesign it completely, to facilitate an easier transition to automation, to separate pedestrians and cyclists from the machines entirely? And then there are the legislative and regulatory challenges, ranging from who or what is at fault in an accident, to how human drivers should interpret the behaviours and intentions of autonomous vehicles when mixing with them on the road. In time, such matters will inevitably be tackled and resolved as and when. Just remember that we went from the first powered flight, 17 December 1903, to the first Moon landing, 20 July 1969, in just under sixty-six years. These giant technological leaps happen, quicker than some expect, and no feat should be deemed impossible.

As you settle into bed, and once you've put down your phone after scrolling through several hundred reels, the clock strikes midnight. In London, over the next twenty-four hours, almost 5.6 million trips will be made by car drivers in the city. Over the course of a year, if you include taxis, around 14.6 billion miles (23.5 billion kilometres) will be covered across the 9,200 miles (148,000 kilometres) of the city's roads. An endlessly churning sea of cars, fuel being burnt, tyres and brakes wearing, batteries depleting, a constantly moving wall of noise, increasing traffic, endless incidents and injuries, jammed streets and car parks, all just to get from one place to another in the most convenient way. Imagine, instead, a world where the mass adoption of the car never came to pass. Perhaps Bertha never met Carl, denying him the support and motivation needed to push ahead with the car, and the public remained unconvinced. Perhaps they continued to enjoy cycling, revelling in the quieter streets, the cleaner air, and leaned towards public transport. Perhaps governments would have realised the opportunities and committed to building accessible and wide-ranging transport networks, shifting the jobs that might have been offered by the automotive world into bus, train, tram, and other self-propelled sectors. A world where the focus was on public, not personal, transportation, creating cleaner, peaceful, more open urban areas and landscapes.

Evidently, this is not where we currently stand. Convenience won out, and the car came out on top. Demand for cars, and car ownership, subsequently continues to rise around the globe, driven by changing economies, environments, and lifestyles. New middle classes are arising, expanding, and striving for that freedom the car is perceived to offer, aided by car-boosting policies, affordable cars, and increasingly widespread and available finance. In the UK, around eighty per cent of new car purchases are made with finance, which in turn has created another swathe of financial and economic problems, all because some buyers feel like they must have a car, the latest car, to keep up with whatever the neighbours happen to be doing, to maintain that supposed symbol of success, irrespective of what they can actually afford. Manufacturing, similarly, continues to escalate and expand, to meet this ever-increasing clamour for cars. But we aren't turfing people off horses, and we can't simply stop those, especially in developing countries, from wanting and embracing what cars can offer. And we can't simply stop people using cars when, in many situations, alternatives don't exist. The challenges, as a result, are significant: more cars, more congestion, more incidents, more stress, more raw materials consumed, more emissions from production and shipping, more fuels burnt, more road development, more green becoming grey, and more debt. While advancing technology, such as automation and ride-sharing, and cleaner electric and electrified vehicles, offers some solutions, coming up with answers to the other problems will take significant time and effort.

That is not to say that cars are without merit, or that they don't have a place in our world. You and I both know how remarkable it is, how much of a privilege and an advantage it is, to be able to step outside, get into a car, and go wherever we want, whenever we want. Because of this, many will tell you that a car is the ultimate symbol of freedom. But, for many more, that freedom is a myth: it's all well and good if you can afford or use it but, if you can't, you're excluded from it. And the existence of the car, and how the world continues to change to facilitate its use, further restricts your personal freedom. If you need a car to go somewhere, if your ability to work is dependent on access to a car, if you have to pour your disposable income into using said car, if you cannot access the services you need without a car, if the car negatively effects the world around you, then that is not freedom: that is a prison. What we must subsequently strive to do in the future, at the very least, is to provide the means for all to travel, by the method of their choice, not just some. Until then, because we must or because we want to, we drive on.

Acknowledgements

My thanks to you, firstly, for taking the time to read this book. I hope that you enjoyed it and that you take something away from it, be it a fact, concept, advice, or inspiration. I would also like to express my endless gratitude to my partner Romana, and to my family, for their continued and unrelenting support as I make my way through the weird and wild land of freelance journalism and writing. Without them, my story would be very different, and far less interesting. And to those, throughout my career, who have provided me with guidance and opportunities, or simply endured my endless chatter about whatever car has caught my attention that day, I tip my hat to you. Lastly, for I otherwise would not be writing this, my thanks to editor Tierney Witty, and the others involved in this project at Seven Dials and Orion Publishing, for their efforts and the opportunity to put this book together in the first place.